When we try to pick out anything by itself,
we find it hitched to everything else
in the universe.

—John Muir

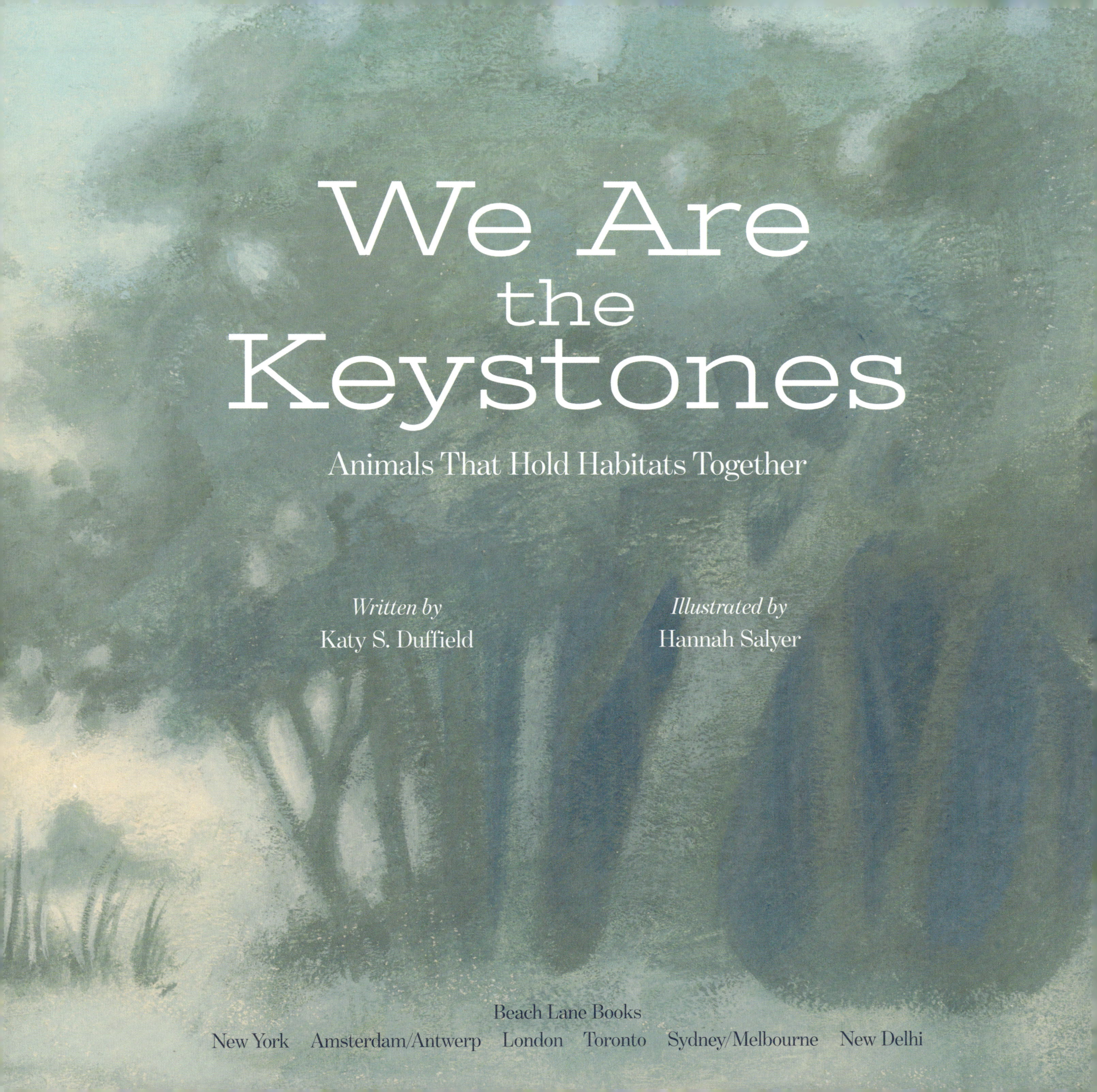

We Are the Keystones

Animals That Hold Habitats Together

Written by
Katy S. Duffield

Illustrated by
Hannah Salyer

Beach Lane Books
New York Amsterdam/Antwerp London Toronto Sydney/Melbourne New Delhi

All throughout the nature-hoods,
in marshes, grasslands, water, woods,

they help themselves—help others too—
just by doing what they do.

They make a difference when they roam,
or when they eat or build a home.

They are the keystones.

Keystone species are plants, animals, or other organisms that have a huge impact on the environments around them. They fill a role no other species can. The loss of a keystone would be disastrous for the plants and animals that depend on them. Without keystones, other species in the community could fail to thrive or even to survive.

We gnaw small trees with our strong teeth,
then tote long limbs that fall beneath.

Construct a dam that slows the flow—
a tranquil wetland starts to grow.

It's a space that suits us all—
the big or tiny, wet or tall.

We are the keystones.

Beaver dams hold back flowing water, creating ponds where beavers build lodge homes. These pond and wetland areas become habitats for turtles, fish, otters, toads, great blue herons, muskrat, salamanders, dragonflies, and many others. Beaver dams also filter and clean the water, producing a healthy environment for all the creatures that live there.

We charge out—a flash of gray!
Then single out, surround our prey.

By thinning herds that overgraze,
we fill our bellies, curb their ways.

Our hunting habits help restore
the aspens, willows, shrubs, and more.

We are the keystones.

Wolves play a pivotal role in controlling elk populations. If too many elk exist in an area, they can eat so many trees, shrubs, and grasses that the land becomes almost bare. This vegetation loss affects birds, insects, and other animals that eat the grass and that live, nest, and shelter in the trees. Fewer trees also mean fewer building materials for the all-important keystone beavers.

We claw our way deep down, around,
create a maze below the ground.

Our tunnel spaces aid air flow,
which helps new vegetation grow.

And that's not all—we *share* our "town."
Need a place? Just come on down!

We are the keystones.

Black-tailed prairie dogs' underground burrows provide shelter and sometimes homes for animals like spiders, snakes, lizards, swift foxes, spadefoot toads, and black-footed ferrets. Their extensive burrow "towns" also allow water and air to move more easily through the ground, which helps plants and grasses to grow—benefiting even more animals.

We take a morning river romp
to snag some salmon—GIANT chomp!

Onshore we dine on skin and brains—
but the carcass, it remains.

The "leftovers"—they fertilize—
prompt new havens to arise.

We are the keystones.

When salmon are plentiful during spawning season, grizzly bears only consume the salmon's fattiest, most nutritious parts (eggs, skin, and brains) and leave the rest of the fish behind. Gulls, ravens, insects, and other creatures make a meal of some of the leftovers while the rest fertilize the soil as it decomposes on riverbanks or forest floors. Richer soil leads to more and stronger trees, grasses, and plants, which birds, insects, and other animals use for homes, food, and shelter.

We visit blossoms, dip our beaks.
Pollen dots our heads, our cheeks.

The teensy grains, we take along;
they keep the flowers growing strong.

More flowers mean more fruits, more seeds—
a gift fulfilling many needs.

We are the keystones.

When hummingbirds carry pollen from one flower to the next, those plants are able to reproduce to make new seeds and fruits that benefit many species of animals and insects. Some keystone hummingbirds, like the female South American green-backed firecrown, have extra-long beaks and tongues that enable them to pollinate flowers that other pollinators can't reach.

With swishing tail, with pointy snout,
we push the marshy muck about.

Groundwater seeps up. (That's the goal.)
We've shaped the perfect gator hole!

Our wallows share a cooling sip,
provide a nice, refreshing dip.

We are the keystones.

During the dry season in the Florida Everglades, water can be hard to find. Gator holes help alligators cool themselves, and these holes also provide creatures such as raccoons, opossums, insects, frogs, fish, wading birds, ducks, and others a respite from the heat, a refreshing drink, and a place to breed or lay eggs.

Fresh-fallen fruit—it's what we eat—
swallowed whole. A nifty feat!

As we digest, we journey out,
leave seed-filled droppings all about.

Then baby seedlings greet the sun,
form future fruits for everyone.

We are the keystones.

The southern cassowary, a large, flightless bird, lives in and around Australia's tropical rainforest. Cassowaries eat the fruits of more than two hundred kinds of trees and can gulp down tennis-ball-size seeds that are too big for other animals to swallow. (They don't even chew—they swallow them whole!) Some rainforest trees depend almost solely on cassowaries to spread their seeds. Without cassowaries, these trees and animals (like flying foxes, fig parrots, caterpillars, and others) might not survive.

With scanning eyes, we glide and seek.
We hunt the sick, the dead, the weak.

We're scavengers—that's what we do.
It sounds disgusting, yet it's true.

But our fine feast promotes sea health.
So, y'all be thankful for our stealth!

We are the keystones.

Tiger sharks, like many other shark species, prey on sick or injured ocean creatures. When sharks eat sick animals, they can keep disease from spreading to healthy marine life. When the weak are weeded out, populations of ocean species become stronger. Some sharks also work as cleanup crews when they eat dead matter that accumulates on the ocean floor.

Our nibbles keep the coral clean.
(That's our day-to-day routine!)

We eat "bad" algae—peck, peck, peck—
to scrape, remove each strangling speck.

With healthy coral by our side,
we *all* can eat and nest and hide.

We are the keystones.

Parrotfish use their odd-looking teeth (called "beaks") to eat harmful algae from coral branches. Without their efforts, coral could die off completely, which would destroy a vital habitat for sea creatures like clownfish, sea turtles, sea stars, seahorses, sharks, eels, and thousands of others. Some estimates note that almost 25 percent of all marine species use coral for their homes.

We float the waves upon our backs
while munching spiky urchin snacks.

Our choice of meal protects the kelp;
the more we chomp, the more we help.

When “forests” flourish, so do we—
along with others in the sea.

We are the keystones.

Pacific Ocean sea otters have one main keystone role: to eat sea urchins. When too many urchins exist, they can consume entire kelp forests, turning the seafloor into a vast, empty space called an "urchin barren." Without kelp, seals, sea stars, baby octopuses, many kinds of fish, and other sea creatures suffer because they lose their source of food and shelter and no longer have a place to lay their eggs.

We use our heads, our trunks, our might
to topple trees; it's quite a sight!

We eat fresh leaves. Bright sun shines down,
which grows more grass—savanna's crown.

Where herbivores can graze anew
(and predators can eat well too).

We are the keystones.

As African elephants travel, they knock down trees, and their big feet tromp down plants. This may sound like a bad thing. But with these plants and trees gone, more sunlight can reach the ground, which helps keep essential savanna grasses growing. These grasses attract and feed grazing animals like Thomson's gazelles and wildebeest—helping meat-eating predators like lions and hyenas find a meal too.

Important creatures. Yes, they are!
Their lives aid others near and far.

Without them, the world would shift.
Their habitats would be adrift.

But when their nature-hoods entwine,
their lives connect . . .
That's when they shine.

They are the keystones.

A Far-Flung Experiment: How Keystones Got Their Name

Robert Paine was a scientist who first visited Mukkaw Bay on the Pacific coast of the United States in the 1960s. As he scrambled along the rocky coastline, he peered into tide pools filled with brightly colored creatures like sea stars, urchins, anemones, sponges, mussels, and barnacles.

One sea star species, the *Pisaster ochraceus*, caught Robert's attention. These large predators, in vivid shades of purple and orange, gobbled up many of the other animals living in the tide pools. This made Robert wonder: What would happen to the tide pool ecosystem if these starfish *weren't* there?

Armed with a crowbar, Robert decided to find out. He began prying the sea stars from the rocks and throwing them far out into the bay.* For five years Robert visited Mukkaw regularly. Each time, he removed all the sea stars from certain tide pools while leaving other tide pools untouched.

Over time Robert discovered something amazing. Tide pools with sea stars stayed pretty normal—about fifteen different species lived there. But tide pools without sea stars became covered in mussels (small shellfish that cling to rocks). Without sea stars to keep mussel populations in control, mussels took over and almost completely pushed out all the other species living in the tide pool.

Robert's experiment showed that *Pisaster ochraceus* played a huge role in its ecosystem. This led him to wonder if other plants or animals would see similar effects on their communities if certain species were removed. To find out, Robert conducted more experiments in other places. In one he looked at different species of sea stars, and in another he studied sea urchins. These experiments confirmed Robert's idea: Some species play larger roles than others in keeping ecosystems balanced and healthy.

Robert called these impactful creatures "keystone species" because they're like the wedge-shaped keystones at the top of building arches. Just as keystones keep arches from falling down, so keystone species keep their communities from falling apart.

Today scientists have identified more than two hundred keystone species of plants and animals.

**Scientists today would likely use gentler methods than throwing sea stars. They might use computer models, lab experiments, temporary barriers, or other ways to study nature without disturbing it.*

Other Keystone Species to Explore

The animals included in this book are only a few examples of keystone species. Many other animals—*and plants*—also play oversize roles in their communities. Here are a few other keystone species to explore:

- Bison
- Woodpeckers
- Honeybees
- Sea turtles
- Coral
- Fig trees
- Saguaro cacti
- Mangrove trees

Selected Sources

Brandt, Laura A., Mark R. Campbell, and Frank J. Mazzotti. "Spatial Distribution of Alligator Holes in the Central Everglades." *Southeastern Naturalist* 9, no. 3 (2010): 487–96. http://www.jstor.org/stable/40985601.

Eisenberg, Cristina. *The Wolf's Tooth: Keystone Predators, Trophic Cascades, and Biodiversity.* Island Press, 2010.

Lindsay, Keith. "What Would Happen If There Were No Elephants?" Journal of African Elephants, February 22, 2019. https://www.africanelephantjournal.com/what-would-happen-if-there-were-no-elephants/.

McConnell, Tatum. "Grizzly Ecology Pt. 3: Bears, Fish, and Trees." Vital Ground Foundation, December 6, 2019. https://www.vitalground.org/grizzly-ecology-bears-fish-salmon-trees/.

Power, Mary E., David Tilman, James A. Estes, Bruce A. Menge, William J. Bond, L. Scott Mills, Gretchen Daily, Juan Carlos Castilla, Jane Lubchenco, Robert T. Paine. "Challenges in the Quest for Keystones: Identifying Keystone Species Is Difficult—but Essential to Understanding How Loss of Species Will Affect Ecosystems." *BioScience* 46, no. 8 (1996): 609–620. https://doi.org/10.2307/1312990.

For More Information

Dunphy, Madeleine. *At Home with the Gopher Tortoise: The Story of a Keystone Species.* Web of Life Children's Books, 2010.

Perdew, Laura. *Keystone Species: Meet the Animals Key to Ecosystem Health and Biodiversity with Hands-On Science Activities for Kids.* Nomad Press, 2024.

"Scientist Profile—Bob Paine." PBS, October 9, 2019. Video, 2 min., 11 sec. https://www.pbs.org/wnet/nature/scientist-profile-bob-paine-mpuvwc/20228/.

"Some Animals Are More Equal than Others: Keystone Species and Trophic Cascades." HHMI BioInteractive, YouTube, May 3, 2016. Video, 19 min., 28 sec. https://www.youtube.com/watch?v=hRGg5it5FMI.

Williams, Lily. *If Sharks Disappeared.* Roaring Brook, 2017.

For all my writer-friend "keystones."
What a difference you've made in my life!
—K. S. D.

For Pat Cummings.
—H. S.

BEACH LANE BOOKS • An imprint of Simon & Schuster Children's Publishing Division • 1230 Avenue of the Americas, New York, New York 10020 • • For information about special discounts for bulk purchases, please contact Simon & Schuster Special Sales at 1-866-506-1949 or business@simonandschuster.com. • Simon & Schuster strongly believes in freedom of expression and stands against censorship in all its forms. For more information, visit BooksBelong.com. • The Simon & Schuster Speakers Bureau can bring authors to your live event. For more information or to book an event, contact the Simon & Schuster Speakers Bureau at 1-866-248-3049 or visit our website at www.simonspeakers.com. • The text for this book was set in Benton. • The illustrations for this book were rendered in gouache paintings and digitally compiled. • Manufactured in China • 0426 SCP • First Edition • 10 9 8 7 6 5 4 3 2 1 • Library of Congress Cataloging-in-Publication Data • Names: Duffield, Katy author | Salyer, Hannah illustrator • Title: We are the keystones / Katy S. Duffield ; illustrated by Hannah Salyer. • Description: First edition. | New York : Beach Lane Books, [2026] | Audience: Ages 4–8 | Audience: Grades 2–3 | Summary: "In this marvelous nonfiction picture book, learn about the hidden heroes of the natural world—keystone species!"—Provided by publisher. • Identifiers: LCCN 2025024388 (print) | LCCN 2025024389 (ebook) | ISBN 9781665946896 hardcover | ISBN 9781665946902 ebook • Subjects: LCSH: Keystone species—Juvenile literature | LCGFT: Picture books • Classification: LCC QH541.15.K48 D84 2026 (print) | LCC QH541.15.K48 (ebook) | DDC 577.2/6—dc23/eng/20251204 • LC record available at https://lccn.loc.gov/2025024388 • LC ebook record available at https://lccn.loc.gov/2025024389